YOUR KNOWLEDGE HAS VALUE

- We will publish your bachelor's and
  master's thesis, essays and papers

- Your own eBook and book -
  sold worldwide in all relevant shops

- Earn money with each sale

Upload your text at www.GRIN.com
and publish for free

Andrea Lieske

Myanmar: pre-colonial & colonial socio-economic developments

GRIN Verlag

Bibliografische Information der Deutschen Nationalbibliothek:

Die Deutsche Bibliothek verzeichnet diese Publikation in der Deutschen National-
bibliografie; detaillierte bibliografische Daten sind im Internet über http://dnb.d-
nb.de/ abrufbar.

Imprint:

Copyright © 2011 GRIN Verlag GmbH
Druck und Bindung: Books on Demand GmbH, Norderstedt Germany
ISBN: 978-3-656-02939-7

This book at GRIN:

http://www.grin.com/en/e-book/180330/myanmar-pre-colonial-colonial-socio-eco-
nomic-developments

Universität zu Köln
Geographisches Institut
Oberseminar: Socio-economic developments in Myanmar

SS 2011

Myanmar: pre-colonial & colonial socio-economic developments

14.07.2011

SS 2011
Andrea Lieske

Studiengang: Geographie (Magister Nebenfach)

Myanmar: Precolonial and colonial socio-economic development

Myanmar: precolonial and colonial socio-economic developments

Table of content

Introduction

1. Geographic & historical basics of socio-economic developments until 1948

2. Overview: The history of Myanmar

3. Pyu City States (ca. 100 – 850 B.C.)

4. Socio-economic developments during the Konbaung Empire

5. Burma under British rule (1852 - 1948)

6. Conclusions: Socio-economic developments in Myanmar until 1948

7. Appendix

8. List of illustrations

9. Bibliography

Introduction

Modern-day Republic of the Union of Myanmar, formerly known as Burma, is shaped by its geographical idiosyncrasies and its history. Both factors are blended into the socio-economic developments of the country which are addressed in this paper. The focus here lies on the pre-colonial and colonial times. The names Burma and Myanmar are used equivalent and without political implications.

To approximate an assessment of those two phases, it is necessary start with a short introduction to the crucial geographic and historical factors affecting the social-economic developments until 1948, the year of Myanmar's independence. The definition of the term socio-economic developments is allocated here as well.

Subsequently a short overview on the most important periods of Myanmar's history is added. The following detailed description of history and socio-economic conditions and developments of three distinct periods, the Pyu city states, the Konbaung Dynasty and the British rule, enables a general view on the socio-economic developments of the time before the state's independence in 1948.

This paper will proceed along one basic assumption: Independent of their time and their technological cultures all three Empires dealt with in this paper had to face the same difficulties: the scarcity of work force due to low density of population and the divide between the 'rice basket' Lower Myanmar and huge areas often depending on its supply of food in the rest of the country. Consequently the history of socio-economic development in Myanmar is hence the history of how each of the examined regimes dealt with those difficulties.

1. Geographic & historical basics of socio-economic developments until 1948

As aforementioned geography and history shaped modern-day country of Myanmar. Among the most formative geographic influences is the general north-southbound direction of mountains ranges, valleys and rivers; as well as the varying quality of soil for agricultural use and the seasonal changes due to the monsoons. Water is the third determinant: Due to the natural barrier for rain clouds of the mountain ranges and the seasonality of the monsoon the availability of water is not always given.

Myanmar: Precolonial and colonial socio-economic development

The development of settlements and later realms was triggered by and adjusted to the geographical conditions. Especially the availability of water and arable soil was indispensable for the establishing of permanent settlements and the change from hunters and gatherers to peasants. Although historical reflection is limited by the number of archeological finds it is possible to distinguish several phases in the more than 750.000 years old history of the country. Myanmar's history until the middle of the twentieth century was coined by three dominant forms of government: city states, dynastic empires and foreign rule in the time of colonialism. The size of the country and its fragmentation into several geographical distinct parts offer an explanation why almost none of the Empires covered the whole area of today's country. Another and possibly even the more crucial factor limiting the possible size of a dominion was population: To control an area a sovereign needs people stationed and living there. Uninhabited country is indeed no man's land. Low density of population marks Myanmar throughout its history. It also indicates why the ruling systems of the different times had different population politics: Some were intended to keep people at one place, others to relocate them somewhere else. Labour force was a scarce resource, thus stable economic development required circumspect population politics often including slavery. To gain surplus labour above the level of pure subsistence was necessary:

> Thus a characteristic of rice was that it could be grown on the same land year after year, without in any fall in yields. Given an adequate water supply, the major determinant of production was labour. (KAUR 2004: 122)

Each period of Myanmar dealt with the labour force problem in an individual way and therefore had characteristic socio-economic developments. The term socio-economic is in this paper used according to the definition by Leser:

> "Sozio-ökonomisch: Bezeichnung für Sachverhalte, Strukturen, Entwicklungen usw., die auf Kräfte, Verhaltensweisen, Aktivitäten und Entwicklungen im Bereich von Wirtschaft und Gesellschaft zurückgehen. [...] Hierbei können die sozialen und die ökonomischen Sachverhalte häufig nicht voneinander getrennt werden." (LESER et al. 2005: 863)

Myanmar: Precolonial and colonial socio-economic development

Transferred to a historical view on Myanmar a number of dominant factors for the socio-economic dynamics manifest themselves: The rice cultivation and its needs; the integration in transnational and even transcontinental trade routes; the low density and inherent mobility of population and last but not least the cohesive religion, the Theravada Buddhism. Based on these conditions different regimes developed, influenced the socio-economic developments of their time and vanished again. The next part of this paper will present a short overview on Mynamar history; followed by an analysis of three main periods in detail. The periods discussed are the Pyu city states, the Konbaung Dynasty and the colonial time under British Rule.

2. Overview: The history of Myanmar

In the context of this paper a historical review in detail is neither possible nor intended. Furthermore the extant findings do not allow conclusions about the socio-economic developments throughout every stage of history.

In general, the history of Myanmar is divided in three main phases. Those are the precolonial times (until 1855), the colonial period (1855 – 1948) and the modern history after the independence in 1948. This paper will focus only on the first two periods. The precolonial times faced a range of migrations and different systems of political organization. The two precolonial systems dealt with in this paper are the Pyu City States and the Konbaung Empire. Those two eras were selected not only due to their importance in the history of Myanmar, but also due to their different approaches to rule: independent, interacting city states (similar to ancient Greek Polis) versus semi-divine centralized Dynasty with hegemonial claims. The annexation by British troops and the transformation into a part of the colony British India marked a caesura of Myanmar's history. Thus the time of the annexation and the decades under British rule and their enormous consequences for society and economy in Myanmar are dealt with as third important period of socio-economic changes.

The prehistory and history of Myanmar is more than 750.000 years old.[1] For the oldest periods finds are rare and thus absolute conclusions are not possible. For later periods more evidences are extant. The different and clearly distinguishable periods of Myanmar precolonial and colonial history are sketched in the table below.

	Time	Period
Prehistory	750.000 – 25.000 BC	Hunters and gatherers
	100.000 – 30.000 BC	Paleolithic occupation
	3.500 – 300 BC	Neolithic Revolution
History	after 100 DC	Raise of Chiefdoms (Pyu States / Mon Kingdoms)
	1044 – 1287	Bagan Dynasty
	1287 – 1531	Small kingdoms
	1531 – 1752	Toungoo Dynasty
	1752 – 1855	Konbaung Dynasty
	1855 – 1942	British rule
	1942 – 1945	Japanese occupation
	1945-1948	British rule
	04.01.1948	Independence

Table 1. Periods of Myanmar history. All dates according to HTIN AUNG (1967) & MOORE 2007

<h3 style="text-align:center">3. Pyu City States (ca. 100 – 850 B.C.)</h3>

3.1. Pyu City States: Background

After the Neolithic Revolution[2], beginning around 3500 B.C., and the associated change of economic behaviour from hunting and gathering to agriculture, permanent settlements

---

[1] *Prehistory* marks the period before first written sources, *history* "begins" with still extant written records. This distinction is here made according to historiographical standards.

[2] The term Neolithic Revolution inclines the dimension of change from hunters and gatherers to a lifestyle based on agriculture and permanent settlement. Please see the definition of Neolitische Revolution by Leser et.al:

> "Neolithische Revolution, Neolithic revolution, in der Jungsteinzeit relative rasch stattfindender Umbruch in der Menschheitsgeschichte, in dem die Grundlagen der höheren Kulturentwicklung gelegt wurden, insbesondere durch sesshafte Lebensweise mit Pflanzenbau und Tierhaltung und die Anlage erster stadtähnlicher Siedlungen." (LESER ET AL. 2005: 606)

evolved in Myanmar. A number of larger settlements already existed when the process of Neolithic change was concluded in 300 AD. These first towns are ascribed to two different cultures, the Mon people and the Pyu people. Those two groups may be described as carriers of the transformation from Bronze-Iron cultures to Hindu-Buddhist walled cities (MOORE 2007: 129). The Pyu, and their city states are portrayed in this part of the paper. The focus is here on the way the Pyu dealt with geographical conditions as well as on the socio-economic developments of the Pyu time.

The question is who were those people whose religious sites and walled cities are still traceable in modern Myanmar? The Pyu people were a lingo-ethnic group which migrated into Ayeyarwady delta in the first century AD. Among several waves of migrants departing from the mountainous region of Eastern Tibet moving southward, the Pyu people are the earliest known wave (LING 1979:5). Traces of them are also found in foreign texts:

> "The first known mention of them is certain Chinese texts of the Tsin Dynasty (265-420CE), which refer to the Pyu as wild and disorderly tribes living in the mountains of China's Burma border; they tattooed themselves, and some were cannibals." (LING 1979: 5)

The Pyu moved into Burma following the course of the river Ayeyarwaddy and came in contact with the Mons and the Buddhism (LING 1979:5). Along their way southwards they founded a number of settlements, some of those later became kingdoms (HTIN AUNG 1967: 7). Their cities existed until the 9th century and were protected by walls including the agriculturally used areas. Their settlements were strategically positioned close to lakes, ponds and streams (MOORE 2007: 10). This was so important because, depending on the monsoon, 80% of the rainfall for some regions occurs in just five months' time. (MOORE 2007: 33).

The Pyu were according to Htin Aung "… more vigorous and more united and thereby able to make the Mon kingdoms to their vassal states." (HTIN AUNG 1967: 8). A clear distinction between Mon and Pyu is on the other hand not always possible. Moore

---

In addition it is necessary to keep in mind that the Neolithic Revolution did not take place at the same time, in the same way or the same speed everywhere.

Myanmar: Precolonial and colonial socio-economic development

suggest to use the terms Austro-Asiatic and Tibeto-Burman groups instead of Mon and Pyu, regarding the two groups not as irreconcilable cultures but as two fragmented language families labeled Mon and Pyu (MOORE 2007: 233). Keeping the difficulties of comparing and distinguishing groups or cultures without sufficient number of sources and finds in mind, I will carry on naming the two ethno-linguistic groups Mon and Pyu in this paper.

3.2. Pyu City states: Socio-economic conditions and developments

The Pyu territory was not marked by one dominant kingdom or a unified empire: The Pyu founded city states which were per se self-sufficient, but interrelated by changing systems of dependencies (HTIN AUNG 1967: 8). They profited from Myanmar connections to transnational trade routes (HTIN AUNG 1967: 7). It is assumable that profits from trade along the overland routes between China and India enabled the formation of the rich and important city states. The favorable location at the mouth of the Ayeyarwady delta enabled maritime trade in addition for the most important city Sri Ksetra. Situated close to today's city of Pyay (formerly Prome) the city state Sri Ksetra was the outstanding Pyu settlement. Economic significance and political dominance allowed the building of religious architecture and strong fortifications as well.

> "By the seventh century the Pyu kingdom of Sri Ksetra [modern: Prome] had become famous in the Buddhist world." (LING 1979: 6)

Sri Ksetra covered an area of around 30km², surrounded by walls of about 15 kilometres length. The name of this largest city of the Pyu means "field of glory" or "auspicious land" (MOORE 2007: 167).

The further question is why were the Pyu so successful in cultivating the land and establishing their cities? The Pyu were able to adapt very well to their environment. Like their Neolithic predecessors they chose to group their settlements close to water sources (MOORE 207:130). Their agriculture and consequently their whole economy were adapted to geographical idiosyncrasies of their new home land. Their success was based on their ecological opportunism.

"These patterns [of change to Hindu-Buddhist walled sites) were largely determined by ecological opportunism in the very different environment found in the north and south of the country. In the arid but fertile lands bordering the Ayeyarwaddy, available streams and in-gyi or lake were maximized to boost cultivation, aided by the incipient local iron production. The production of brick, for buildings and walls, was an essential corollary in the digging of shallow moats buttressing the walls, the water in resulting depressions augmenting the crop yield while demarcating site domain." (MOORE 2007: 130)

The flourishing of the Pyu cities and their trade was enabled by their agricultural expertise. The Pyu were integrated into transnational trade routes, at the latest after the increase of trade with India after the first century. [3] Wet rice cultivation[4] in the favorable delta region even enabled a surplus extraction and hence trade with the staple food. Like in other Pyu cities, Sri Ksetra's walls did not only include palaces, houses and religious buildings like pagodas and monasteries but also large agricultural area. This fact illustrates both the importance of rice and the permanent danger of war. A common reason for war was the acquisition of labour force, which was the limiting factor of Pyu economy. The common consequence of defeat was the enslavement of the people of the population. The Pyu states' economic power was defined and limited alike by their control over people. The low density of population made humans a valuable good. War slavery and peonage due to crime or debt were common (HTIN AUNG 1967: 8). Therefore the architecture of the walled city states indicates the constant danger of war, the necessity of protection by walls and the ability to resist a long siege by including cultivated land. Need for labour force marked the whole region and made slavery a common consequence of war. The importance of the labour force created by enslavement and the special status of the slave is described by Kaur:

---

[3] The trade between India and Myanmar was increased after the Roman Empire cut the eastern trade due to Vespasian's prohibition to export gold. Please see Aung: "… a great expansion of trade between the two regions [India and Roman Empire] occurred. As the balance of trade was so much in favour of India there was a great drain of gold from the Roman Empire, and Emperor Vespasian (A.D. 69-79) was constrained to prohibit the export of gold from his dominions; India was then forced to turn to Southeast Asia for a new source of gold. This gold crisis coincided with the great advance in navigation, as a consequence of the 'discovery' of the monsoon winds by 'outsiders'." (AUNG 2002:9)

[4] Concerning rice: The rice plant *Oryza sativa* is not a water plant, but needs high and constant water supply. Therefore irrigation works are often necessary. Labour is the crucial factor for surplus extraction.

Myanmar: Precolonial and colonial socio-economic development

"Slavery was inevitably a consequence of war, which made available a group of prisoners for the labour pool. Since it originated in an act of violence its continuance required coercion. The slave was seen as an outsider, as property that had salability value." (KAUR 21)

The Pyu were by fare not the only slaveholding society of the region. The scarcity of humans and the consequent lack of labour force was a supraregional problem. After armed conflicts with the Nanchao Dynasty in the 9th century many of the Pyus fell into slavery themselves (LING 1979: 8). [5]

The enforcements of the city states were thus on the one hand created to protect their wealth and their people. On the other hand they enabled a strict control over the movement of the most important resource, the humans. Although many aspects of socio-economic conditions in Pyu city states are still opaque, it is possible to conclude a social stratification. The Pyu society was vertically stratified in different classes; namely gentry, clerics, military personnel, craftsmen and peasants. Slaves were as aforementioned important part of the labour force, especially for the agricultural and the construction sector. The immense harems of the kings (HARVEY 1967: 13) reflect the prevalence of polygamy in Pyu society as well as the importance of women as status symbols. All in all the relevance of rice cultivation, trade and control over people are outlined already in these societies which existed from the first to the ninth century. The humane law and the tendency to solve conflicts by negotiations, royal marriages or agent's combats (HARVEY 1967: 12) probably show the wealth of the kingdoms, the pacifying influence of highly respected Buddhist monks and once more the high value of humans as economical resource.

While their Neolithic predecessors inhabited many river valleys and are still traceable merely by their stone and bronze-iron artifacts mainly along the middle and upper Ayeyarwaddy, the Pyu were among the first humans to shape the landscape of Mynamar permanently. The remains of their giant city walls and their sophisticated irrigation system are detectable in today's landscape without too much trouble. Their ecological

---

[5] This loss of war, the conquest of their cities and the abduction of large parts of the population marked the end of their eight century long reign.

Myanmar: Precolonial and colonial socio-economic development

opportunism shows in the choice of places for settlement and fields. Thus it is possible for Moore to conclude: "Major alteration of the landscape begins with the emergence of walled sites (circa 200 BC – 900AD)." (MOORE 2007: 10).

All in all the power base of the Pyu city states was their control over people by walls, wars and slavery and as well their ecological opportunism resulting in sophisticated agriculture. As assumed before, the Pyu faced the problem of scarcity of labourers. But the Pyu were not forced to find a way to feed less favorable agricultural regions. Their brick walled cities offered sufficient room and cultivation area for their people. On the other hand, the growth of the population was limited by the enforcements as well. The Pyu city states were and remained self-sufficient until their fall.

4. Socio-economic developments during the Konbaung Empire

4.1. Konbaung Empire: Background

The Konbaung Empire arose after the fall of the formerly reigning Toungoo Dynasty in 1752. The founder of the Dynasty, Alaungpaya, who was initially a common man, managed to capitalize on the chaos after the state collapse. He built a large Empire, undisputed at the latest after the conquest of Bago city in 1759.

Roughly thousand years after the end of the Pyu city states the dominant socio-economic factors had not altered very much: rice cultivation, low density of population, Buddhism. But one important factor was indeed different and brought along change in all socio-economic fields: The Empire consisted of many cities and a large area. Moreover its capital and thus the centre of the Empire were not in the fertile delta of the Ayeyarwady. Instead the different capitals of the Konbaung Dynasty were further north, the last of them in Mandalay. The agricultural conditions were much less favorable in this part and the principal region could not survive without food imports from the south.

4.2 Konbaung Dynasty: Socio-economic conditions and developments

The immense demand of food for the more populated northern city regions characterized the politics of the Konbaung Dynasty and thus the socio-economic developments. Hence Alaungpaya and his descendants did their utmost to turn the Ayeyarwady delta and

Myanmar: Precolonial and colonial socio-economic development

Lower Burma[6] into the Empire's food base. After the disempowerment of local rivals, the preconditions were good:

"Lower Burma was flat, fertile, and very sparsely populated; recent estimates put the population density at only 10 persons per square kilometer. Most of the land was uncultivated." (VAN SCHENDEL 1991: 54)

To improve the surplus extraction large resettlements were initiated by the governments. The migration of ethnic Burmese originated in Upper Burma towards arable land in Lower Burma was encouraged and sometimes coerced (KOENIG 1990: 13). Parallel to the southward movement of Burmese a Burmanisation of the southern areas occurred.

"A major development of the Konbaung period was the gradual 'Burmanisation' of Lower Burma. Migrants from Upper Burma in search of land on which to settle came in such numbers that Burmans began to predominate numerically in some parts of the delta. This dampened local resistance against Konbaung rule, which had been fuelled partly by ethnic differences between Burman overlords and Mon subjects." (VAN SCHENDEL 1991: 57 - 58)

The Konbaung Dynasty was in contrast to other Myanmar kingdoms very inward-looking. Isolation of the country was a consequence of their seizing power and a principle of their policy. They cut off the land from almost all trade; the southern part of Myanmar was affected by this politic to greatest extent.

"After the 1750s, the Konbaung state had cut off Burma off from most of the Asian maritime trade network by proclaiming a royal monopoly on external trade which implied a ban on the export of rice and many other goods.[....] Earlier incentives for producing rice surpluses in Lower Burma vanished and its trading communities and market towns went into decline. By early nineteenths century the region had been reduced to an undeveloped backwater..." (VAN SCHENDEL 1991: 54)

---

[6] In this part of the paper I use the word Burma as in "Lower Burma" to describe a region. No political intention is included in this nomination.

Myanmar: Precolonial and colonial socio-economic development

The new reign, the trade stop and the demand for rice shaped the economy and social life in Lower Burma in a new way. The economy was after the decline of trade restricted to subsistence and barter. The surplus extraction was now destined for the benefit of the faraway court and was organized by means of a new class of officials:

> "The central were presented locally by district governors (*myowun*), who were supposed to rule over a number of *myothugyis*, or headmen of 'townships' consisting of several villages. In reality most power was in the hand of these lower officials. *Myoththugyis* were small territorial lords who acted as state official of the lowest level." (VAN SCHENDEL 1991: 53-54)

In fact, power was in the hand of the *Myoththugyis* who acted as brokers between peasants and state. Their office was hereditary and enabled them to collect up to two-thirds of the surplus (VAN SCHENDEL 1991: 56). Nevertheless their subordinates were not completely at their mercy: Land was assessed as a free gift of nature and available in abundance. If necessary, the people were mobile and free to move.

> "Lower Burma's decommercialised, de-urbanised society showed the fluidity and mobility usually associated with frontier societies. Both geographical and social mobility were common and, despite sumptuary laws, social differentiation was far less marked than in Upper Burma. "(VAN SCHENDEL 1991: 58)

Hence the peasants could move on to another area of land without much difficulty, because there was no general peonage or restriction to a specified area. Apart from this freedom of movement, life in Lower Burma was tough for common people. The economy descended, the cities and market towns sank into insignificance and the formerly members of non-peasant classes like craftsmen emigrated or became cultivators. The previously prosperous coinage declined along with the economic strength (KOENIG 1990: 14).

> "Economic activity became restricted to subsistence-orientated rice cultivation, fisheries, and the manufacture of salt. Although there was some circulation of coins and bits in the coastal regions, and lead, copper, and silver were known as currency, then inhabitants used to exchange rice for the few consumer goods they needed: cotton yarn, pottery, simple agricultural implements, and draught animals." (VAN SCHENDEL 1991: 54)

By the export ban on rice and other agricultural goods of the new regime, the southern part of the Empire was cut off from international trade. The northern part meanwhile established trade relationships with European powers. Especially the trade with weapons and above all ammunition provided by British and French trade companies flourished. (KOENIG 1990: 25)

In the south, Lower Burma regressed into a purely agricultural society. Often less rigidly regulated and with plenty of arable land, Lower Burma was farmed mostly rather extensive than land-saving. There was as good as no competition for good farm land beyond the reserved areas for monasteries. Thus it was easy for commoners to acquire land possession:

> "The first cultivator to clear and cultivate a field in Lower Burma obtained an informal tenure right known as *damaugya* and if the family continued in possession for three generations the land would become 'ancestral property' (*bobabaing*)." (VAN SCHENDEL 1991: 57)

Human labour force was the crucial factor in surplus generating in the wet rice cultivation of Konbaung times. Still wage labour was of no importance. Family labour was the standard and potential scarcity of workers was equilibrated by labour exchange between households. (VAN SCHENDEL 1991: 57)

Although peasants were free to move in the southern region without severe political limitations, the Konbaung developed a complicated work distribution system. The lowest class of workers was the slaves, many of them owned and employed by religious institutions.

"In Burma [...] there were monastery slaves who lived in villages as ordinary peasants. They gave produce of their assigned for the upkeep of Buddhist religious establishments." (KAUR 2004: 21)

While slavery was usually a consequence of war and involved abduction and coercion of foreigners, subjects of the Konbaung kingdom could also fall into slavery due to indebtness. The resulting debt slavery or debt bondness could also be hereditary. While the bonded person could theoretically return to the class of free men after repaying the debt by labour, he or she was until then indeed not free:

"Bondage was also transferable and even saleable, and in practice the right to redemption was little more than nominal." (KAUR 2004:22)

But not only slaves and debt slaves worked without payment. Public projects like the building of infrastructure often were based on corvée.

"Corvée was an obligation of the subject or peasant class to the ruling elite. In many areas, people had an obligation to labour for the state or its representatives that could last up to six months at a time." (KAUR 2004: 22).

It is assumable that corvée work substitutes for taxes in a kingdom where money was much less scarce than work force. The differentiated systems to oblige humans into unpaid work and to control their movements alike show the importance of labour force and its organization for the last great kingdom in Myanmar.

Indeed the Konbaung government faced the same difficulties as other regimes in Myanmar before: water supply difficulties, scarcity of labour force, differences in agricultural productivity. The largeness of their Empire and the north-south divide of population and agrarian production increased the problems. Higher density of population in the north and simultaneous weaker productiveness of farming there created demand for food supply from the fertile southern part of the country. The Konbaungs' answer to the food scarcity was the cut of all food-related export and thus the metamorphosis of Lower Burma in a food basket neglected in all other aspects. The sophisticated irrigation systems developed to enable surplus extraction from farming in the northern part and the trade systems were oriented completely to support the northern part of the reign, including the capital Mandalay, could nonetheless not prevent famines. The Konbaungs' centralized country was weak at its frontiers: Military, economic and human power was concentrated on the all-important capital region. The existing problems increased in the last decades of the Empire after the loss of Lower Burma and as such the main food source during the 2nd Anglo-Burmese war of 1852. Moreover many young men migrated to the now British ruled south to work and eventually to live there (VAN SCHENDEL 1991: 88). This intensified the labour force problem and the food shortage in the north. The fall of the northern capital region and the Konbaung Empire in the 3rd Anglo-Burmese war in 1885 was the consequence of the British aggression and the home made problems of the slowly collapsing state of the Konbaung Dynasty.

5. Burma under British rule (1852 - 1948)

5.1. Burma under British rule: Background

Myanmar was part of international trade since at least the first century A.D. This trade intensified with the appearance of the Europeans in Asia. After the Portuguese Vasco da Gama discovered the direct shipping route to India and thus to Asia, the region became part of the conflicts for power and economic dominance of the European powers. For Burma this ended in three wars against British troops in the nineteenth century. Even centuries before the British had like other European powers intense trade relations to South East Asia, mainly executed by powerful trade organizations like the English East India Company. [7] The British Empire was founded mainly on sea trade and colonialism. In Burma the British employed the same politics. The 2nd of these Anglo-Burmese wars ended with British victory and the annexation of Lower Burma by British troops in 1852.

---

[7] Please see Kaur 2004: "Southeast Asia was integrated into the world trade system between the fifteenth and seventeenth century and came to play and increasingly important role in intra-Asian trade. A dominant feature of the period was the existence of European maritime or seaborne Empires based on the control of sea routes. European power on land was relatively limited and centered on fortified trading centers such as Melaka and Batavia. Trade was organized through chartered companies, the most prominent being the English East India Company and the Dutch East India Company. These companies were largely independent of home governments, had the power to conclude treaties with local rulers, and possessed military forces to protect themselves." (KAUR 2004: 5)

Myanmar: Precolonial and colonial socio-economic development

The remaining Konbaung Empire lasted only three decades longer, before it was conquered by British troops during the third Anglo-Burmese war in 1885. In 1886 Burma became officially a part of the colony British India. Burma's fall under foreign influence was not unique, the conquest by the west and the following economic domination were common features of the South East Asian history between 1870 and 1940 (KAUR 2004: 6). The British reign lasted until the independence of Burma in 1948, only interrupted by the conquest and occupation of the country by Japanese troops during the 2nd world war. The five decades of British reign changed the country's economic and social structure fundamentally.

General historical background	Date	Development in Burma
Colonialism	1824	1st Anglo-Burmese war
	1852	2nd Anglo-Burmese war, British conquest of Lower Burma
	1870 -1890	Open frontier
	1885	3rd Anglo-Burmese war, British conquest of Upper Burma
	1886	Burma is declared a British colony
	1886 - 1900	Guerilla war against British
	1990 - 1919	Maturity & change of rice industry
Global finance crisis after stock market collapse (Black Friday)	1930 - 1940	Social problems
	1930 - 1932	Peasant rebellion a.ka. Hyasa San Rebellion
2nd world war	1937	Burma becomes separated from India
	1942 - 1945	Japanese occupation
	1945 - 1948	British rule again
End of colonialism	1948	Independence

Table 2. Overview of developments in the history of Burma in Colonial times (1824 -1948). All dates according to HTIN AUNG (1967), KÖNIG (1990), AUNG-THWIN (2002) and VAN SCHENDEL (1991)

5.2 Burma under British rule: Socio-economic conditions and developments

After the annexation of Lower Burma in 1852 the Konbaung Empire was cut off from their main food supply; in the south the British government was in charge of the development and exploitation of the newly acquired possession. The colonial masters' primary aim was economic success. The exploitation of Burma's ore was one way to it; the other was the use of the enormous agricultural potential of Lower Burma. As aforementioned arable land existed in abundance, but the lack of work force was crucial once more. Furthermore the former Konbaung regime failed to supply the necessary infrastructures and especially failed to control the *myothugyis* to avoid corruption and despotism. The British solved these difficulties by providing the necessary infrastructures and took the power from the *myothugyis,* thus reducing them to minor servants (VAN SCHENDEK 1991: 86).

To enable maximal economic exploitation of the new area, the British government invested massively in Burma. The increase in Foreign Direct Investments (FDI) was the engine to productivity and is exemplary depicted in the Table below for the years 1914 - 1937.

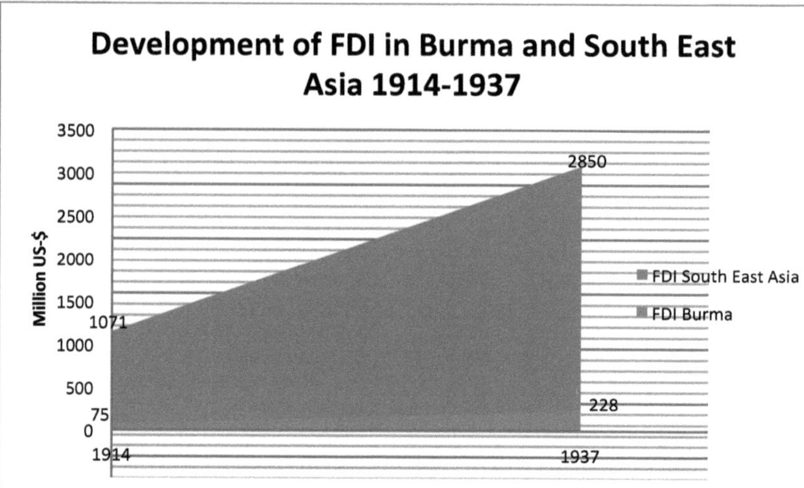

Table 3. Development of Foreign Direct Investments (FDI) in Burma and South East Asia. All data according to Kaur (KAUR 2004: 29)

The British Foreign Direct Investments were by fare not only intended for the agricultural sector. The biggest part of the FDI was invested into Mining, which is by far the most capital-intensive sector of the four sectors pictured in the table below. The amount of 4% of the FDI for "Other" sectors is an indicator for the (successful) attempt of the British administration to ameliorate the irrigation system and infrastructure as well.

Myanmar: Precolonial and colonial socio-economic development

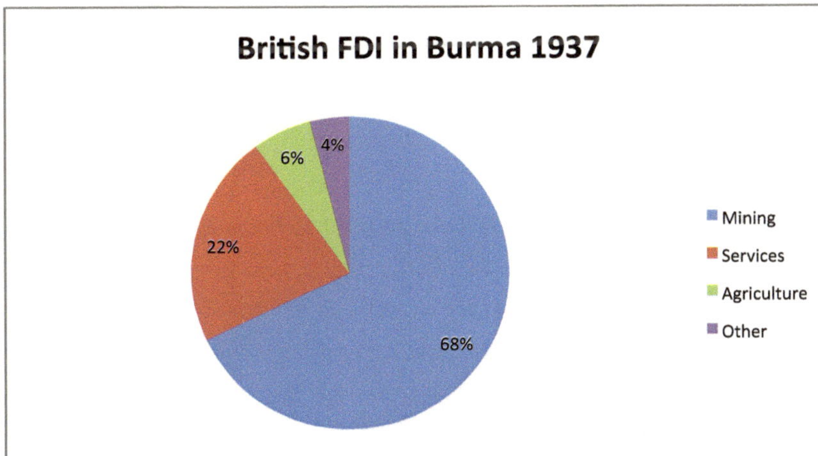

Table 4. British Foreign Direct Investment (FDI) in Burma 1937. All data according to Kaur (KAUR 2004: 33)

The main problem for the Burmese economy had never been the need for investments alone; it was the lack of labour force. The prosperity of economy in British time increased the need for workers even more. A problem of the region existing since thousands of years was now the major obstacle for the success of the eastern part of the colony British India. The British chose measures formerly unknown to the region and its population to overcome this impediment for the exploitation of the country. They encouraged migration into Lower Burma from Upper Burma and India, seasonal as well as permanent. Wage labour became the dominant way of labour during their reign. (KAUR 2004:125)

The socio-economic developments of the country under British rule show best at the example of sector where the majority of workers was employed, the agricultural industry. The main crop was still rice; the staple food became important for the food supply of the whole British Empire. It is possible to distinguish phases of development of the Burmese rice industry:

"Three phases characterized the Burmese rice industry: the open frontier 1870-1900; maturity and change 1900-1929; and depression and social problems 1930-1940." (KAUR 2004: 125)

Myanmar: Precolonial and colonial socio-economic development

During the time of the so-called "open frontier" in 1870 -1900 inner-Burmese migration was the source of the much needed labour force. The British encouraged the inhabitants of Upper Burma, also of the Konbaung Empire, to migrate into the South with well paid work and tax exemptions.

> "But other government measures were designed to encourage migration and extension of the cultivated areas: newly occupied land was exempted from taxation for a period varying from one to twelve years, rates on fallow land were kept very low, and immigrants from Upper Burma did not have to pay a minor tax (the capitation or household tax) during their first two years in British Lower Burma." (VAN SCHENDEL 1991: 88)

For the first phase of the rice industry of Burma the migrants from Upper Burma were sufficient to feed the work market. Moreover the aforementioned structures of work distribution during the Konbaung time, especially corvée and slavery were not abolished by the new rulers at once. Indeed was the abolition of slavery only concluded in the 1920ies (KAUR 2004:23). But in general terms, employment was now based on wage labour, creating a need for money. This encouraged the first wave of migrants from India.

> "During this period therefore, the Delta peasant cultivators relied on Burmese workers, who received payment in cash, to clear the land and prepare the fields. [...] activities were carried out before the harvest, workers needed payment in cash. From about 1880, Chettiar moneylenders from India arrived in the delta region to provide loans to cultivators. Chettiar credit and rural banking services were thus essential to the expansion of the industry." (KAUR 2004: 125)

The rice industry was still dominated by the independent Burmese rice producer in the decades until 1890, although the Chettiars played an important role as moneylenders and the Upper Burmese migrants as seasonal workers. Nonetheless the opening of the work market, thus the competition for labourers and the investment of the British were successful in creating a tradable surplus. The area used for rice cultivation increased many times over, as depicted below.

Myanmar: Precolonial and colonial socio-economic development

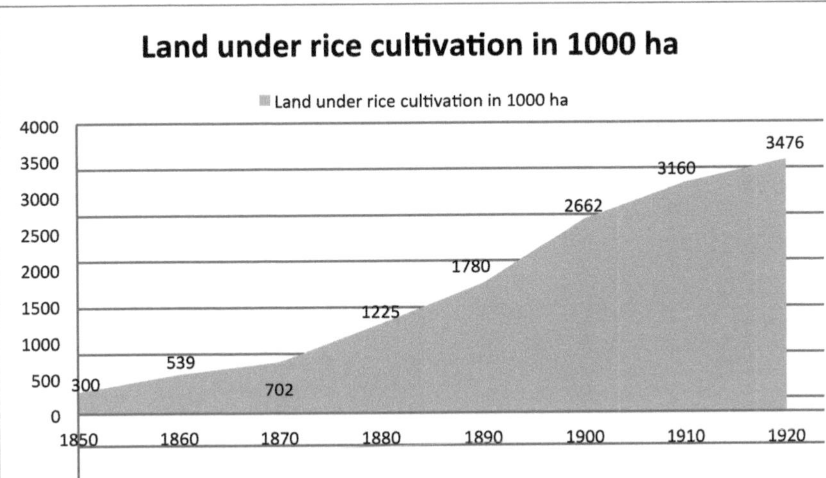

Table 5. Lower Burma, land under rice cultivation in 1000 ha, 1850 - 1920. All data according to Kaur (KAUR 2004: 124)

This great success in intensifying the rice production was important for the British; the staple food was needed in their Empire.

> "Demand for rice tended to rise rapidly as the nineteenth century progressed. The export demand for Burmese rice came initially from India, then from Europe, and then increasingly from other Asian markets." (KAUR 2004: 124)

The export of rice from the Burmese delta multiplied and it was soon once more the most important cash crop. Governmental schemes to introduce other cash crops like coffee and cotton were wrecked by the high rice prices. The price increased even more when loss of production occurred, for example during the Indian strikes in the 1860s or when the American civil war stopped the intense rice production in the two Carolinas (VAN SCHENDEL 1991: 87). The trade with rice was mainly in the hand of European merchants, but the Burmese cultivators, middlemen and workers also profited from the development on the international food market.

Myanmar: Precolonial and colonial socio-economic development

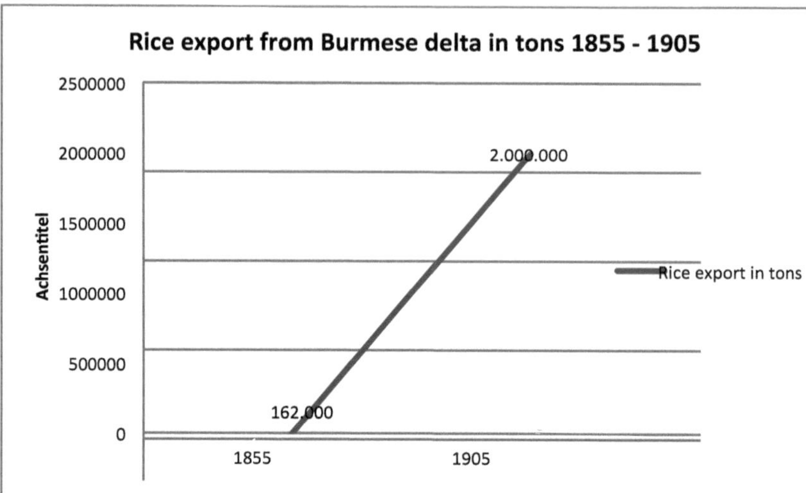

Table 6. Rice export of Burmese delta in tons. All data according to Kaur (KAUR 2004: 125)

The success of rice cultivation and export changed the society. In the first decades the life standard increased for the majority of the people. The major changes developed especially in the 2nd phase of the rice industry from 1900 -1930. During this period of full maturity of the rice industry, the agricultural sector changed and became an industrial agriculture, where wage labourers were hired for single tasks like ploughing or planting. The economic development was so intense that labour force was once more scarce. More immigration and a new division of labour were the consequences.

> "During this period too, seasonal Burmese labour had become scarce and Burmese cultivators turned to Indian migrant labour. [...]The migrants came principally as free labour, hired either by recruiting firms or by crew bosses known as maistry. The maistry organized the labour gangs that moved through the rice districts, contracting and completing jobs before moving on to a new rice district." (KAUR 2004: 126)

The Indian workers were only employed seasonal from September to march. In the meantime they sought work in the cities or returned to India. There number fluctuated therefore very much, but it is possible to estimate it as around 284.00 in 1900 and

Myanmar: Precolonial and colonial socio-economic development

777.000 in 1930 (KAUR 2004: 126). Their importance for the rice industry and the new division of labour are depicted in the table below.

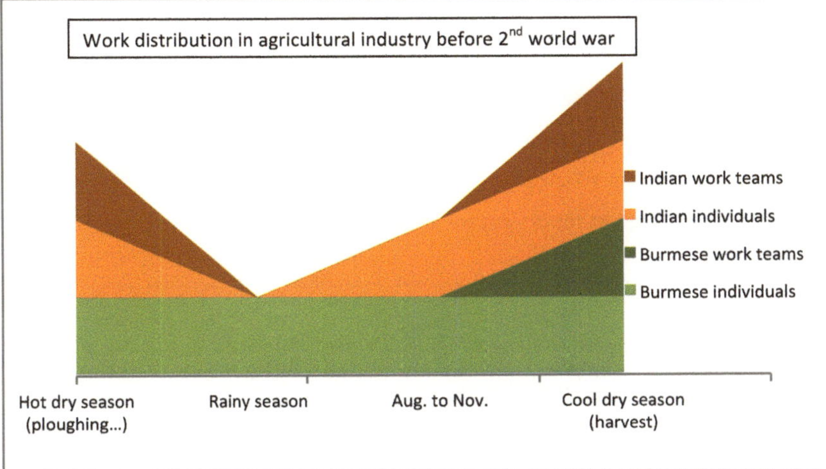

Table 7. Work distribution in agriculture and industrial agriculture. All data according to Kaur (KAUR 2004: 132)

The new division of labour included a new stratification in the society. The new money-oriented economy developed very well regarding the moneylenders and the British colonial masters. But due to the turn towards wage labour and especially the turn towards land as sellable property of the state and subsequent land taxation many commoners impoverished. Rising land alienation and indebtness among Burmese cultivators triggered xenophobic tendencies in the Burmese population. (VAN SCHENDEL 1991: 86)

"The nationalist movement, which combined anti-colonial sentiments with a concern to establish a more equal society, now became a vehicle for the impoverished 'sons of the soil' in both urban and rural areas. Burma was not the only part of British India in which large numbers of outsiders were introduced by colonial overlords. This makes the development of Burmese nationalism all the more significant. Burma was the only part of the colony in which anti-foreign and

Myanmar: Precolonial and colonial socio-economic development

especially anti-Indian sentiments were so intense that the Indian nationalist movement proved powerless." (VAN SCHENDEL 1991: 138

The tensions erupted when the global finance crisis after the stock market crash of the Black Friday (24.10.1929) hit Burma. The sudden collapse of the rice price led to further indebtness and decline of most rural incomes. The brittleness of the social fabric was exposed, the conflict between the poor and the powerful ignited. Riots occurred first in the capital Rangoon (today`s Yangon) and then everywhere in the country. The Indian migrants were the scapegoats for the miserable situation of the Burmese and were often the first to be attacked.

> "The first riots occurred in Rangoon in 1930; they were followed by many rural insurgencies that have become collectively known as the Hsaya San Rebellion or the Peasants' Revolt of 1930-32. [...]It consisted of organized attacks on village headsman, Europeans, railways, police stations, telegraphic equipment, and even military columns; communal assault by Burmans on Indians and Chinese; and dispersed guerilla resistance and banditry. The revolt was a political reaction to the economic crisis and was possible only because of the well established network of nationalist organizations at the village level." (VAN SCHENDEL 1991: 139)

The revolt was the most extreme symptom of the 3rd phase of Burmese rice industry, the decade of social problems according to Kaur. The life standard declined for the mass of the people, poverty, indebtness and even hunger were the consequences of the global market crash. Nonetheless, while the majority of Burmese people worked in the agricultural sector and suffered from the crisis, the economy all in all was not hit this hard. The rice price was down, but other prices were climbing. This was partly due to the military needs of the 2nd world war, beginning in 1936. The production of other commodity goods like rubber remained high. [8]

The stock market crash ignited the problems in British ruled Burma, but they were indeed home-made by the British rule. While the encouragement of migration was the solution

---

[8] Please see tables 8 and 9 (appendix) for details on the production of commodity goods in Burma at the beginning of the 2nd world war.

Myanmar: Precolonial and colonial socio-economic development

for the labour force problem and thus the key to greater economic success, this measure was on the other hand problematic for the inner peace of the country. The felt inequality in comparison to the colonial masters and the new migrants from India, the rising prices for land and commodities of daily life, the problem of indebtness and finally the economic crash led to the xenophobic riots. These riots were not only directed against Chettiars and other Asian foreigners, but were indeed the beginning of the fight for independence of the South East Asian country.

The British managed to tackle the main problems for the socio-economic developments (work force, infrastructure especially water supply and supply of food for all parts of the country) only at the cost of creating new problems. The social tensions their political and economic measures created were finally the end of the British rule over Burma. In 1937 the country was separated from the colony British India, in a vain attempt to calm down the rebels by changing the name disapproved by the nationalists. After the Japanese occupation (1942 -1945) the British gained control once more, but were unable to resist the Burmese freedom movement for long. In 1948 the country Burma declared its independence. The Burmese even left the Commonwealth of Nations, as only one of the former British colonies. This illustrates the rift between former colonial power and its former subjects.

6. Conclusions: Socio-economic developments in Myanmar until 1948

This paper was based on the assumption that the rulers through all the time, from precolonial Pyu kings to the British government in the time of Colonialism, faced the same difficulties in establishing successful economy and rule in Myanmar. This assumption was partly correct: All three examples here examined more closely had to face the problem of low density of population and consequent demand for more labour force. The divide between the 'rice basket' Lower Myanmar and huge areas depending on its supply of food in the rest of the country was a challenge for the administration through all times as well. Notable exceptions were the Pyu kingdoms whose small reigns did not expand into non favorable regions. Indeed it is possible to conclude, the history of socio-economic development in Myanmar is hence the history of how each of the regimes dealt with those difficulties.

Myanmar: Precolonial and colonial socio-economic development

The Pyu relied on their high walled cities to confine their own population and slavery to supply them with more workers. Their irrigation system and cultivation was sophisticated. Their success was probably also due to the small number of inhabitants in a city. Their self-sufficient system worked well for eight hundred years, but in the end they were conquered and enslaved themselves by other kingdoms.

The Konbaung Empire was also build upon slavery, plus a sophisticated system of corvée. They did not like the Pyu before restrict themselves to one city or area. In the end their country was indeed too large: The northern part around the capital Mandalay depended on the food supply from Lower Burma. The Konbaung made the former prospering Lower Burma a pure rice basket, only existing to provide food for the north. This centralization on the capital made their Empire weak at its frontiers. When they lost Lower Burma and as such their main food source their Empire declined. The population density was too high and the agricultural productivity too low to maintain the kingdom. Famines and finally the conquest by the British were the result of the Konbaungs' one-sided policy of isolation.

The British solved the demand for workers by encouraging immigration and the introduction of wage labour as main kind of employment. In this way and thanks to high Foreign Direct Investments they succeeded in increasing the rice cultivation, the export of the crop and the whole economy enormously. Under the British dominated the competition for wage labourers and prices, always under conditions of a free market economy. But their solution for the scarcity of work force led to social tensions in the country which ignited after the global finance crisis in 1929/ 1939. In the end the British had to give up their colony Burma.

The closer look at the socio-economic developments during the precolonial and colonial time makes it possible to conclude that all ruling systems did indeed face similar problems and also used individual approaches in their attempt to find solutions to boost their economy. The details of these solutions coined the socio-economic developments in Myanmar of their time.

7. Appendix

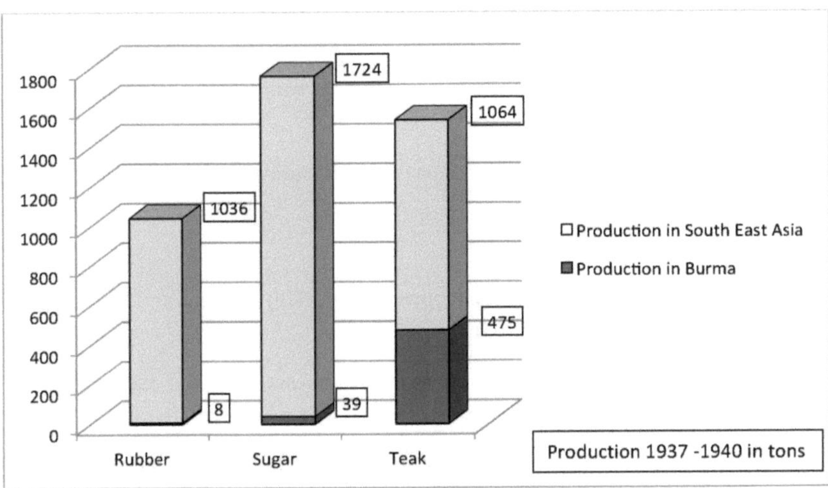

Table 8. Production of rubber, sugar and teak in South East Asia and Burma in 1937 – 1940. All dates according to Kaur (KAUR 2004: 21)

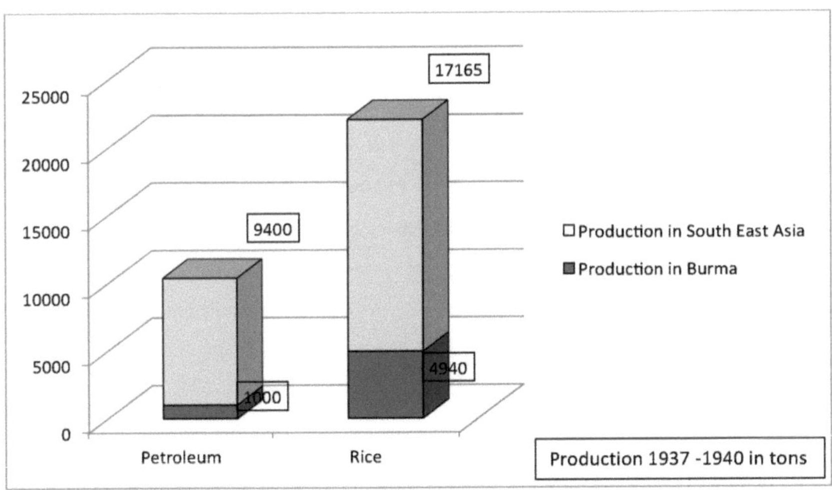

Table 9. Production of petroleum and rice in South East Asia and Burma in 1937 – 1940. All dates according to Kaur (KAUR 2004: 21)

Myanmar: Precolonial and colonial socio-economic development

8. List of illustrations

- Table 1. Periods of Myanmar history. All dates according to HTIN AUNG (1967) & MOORE 2007

- Table 2. Overview of developments in the history of Burma in Colonial times (1824 -1948). All dates according to HTIN AUNG (1967), KÖNIG (1990), AUNG-THWIN (2002) and VAN SCHENDEL (1991)

- Table 3. Development of Foreign Direct Investments (FDI) in Burma and South East Asia. All data according to Kaur (KAUR 2004: 29)

- Table 4. British Foreign Direct Investment (FDI) in Burma 1937. All data according to Kaur (KAUR 2004: 33)

- Table 5. Lower Burma, land under rice cultivation in 1000 ha, 1850 - 1920. All data according to Kaur (KAUR 2004: 124)

- Table 6. Rice export of Burmese delta in tons. All data according to Kaur (KAUR 2004: 125)

- Table 7. Work distribution in agriculture and industrial agriculture. All data according to Kaur (KAUR 2004: 132)

- Table 8. Production of rubber, sugar and teak in South East Asia and Burma in 1937 – 1940. All dates according to Kaur (KAUR 2004: 21)

- Table 9. Production of petroleum and rice in South East Asia and Burma in 1937 – 1940. All dates according to Kaur (KAUR 2004: 21)

All illustrations used in this paper are developed by the author. The authors providing the data are mentioned below each illustration and in the list of illustrations.

9. Bibliography

AUNG-THWIN, M. (2002): Lower Burma and Bago in the History of Burma. – In: Gommans, J.: The maritime frontier of Burma. Exploring Political, Cultural and Commercial Interaction in the Indian Ocean World 1200 – 1800. Leiden. 25 – 53.

HARVEY, G.E. (1967): History of Burma. From the Earliest Times to 10 March 1824. The Beginning of the English Conquest. – London.

HTIN AUNG, M. (1967): A history of Burma. – New York.

KAUR, A. (2004): Wage Labour in Southeast Asia since 1840.Globalisation, the International Division of Labour and Labour Transformations. – Armidale.

KOENIG, W.J. (1990): The Burmese Polity, 1752 – 1819. Politics, Administration, and Social Organization in the Early Kon-baung Period. – Ann Arbour.

LESER, H. (Hrsg.) (2005): Wörterbuch Allgemeine Geographie. Aktualisierte Neuausgabe. – München.

LING, T. (1979): Buddhism, Imperialism and War. Burma and Thailand in modern history. - London.

MOORE, E.H. (2007: Early landscapes of Myanmar. – Bangkok.

VAN SCHENDEL, W. (1991): Three Deltas. Accumulation and Poverty in Rural Burma, Bengal and South India. –London.